EMMANUEL JOSEPH

Temples of the Flesh, Exploring the Mythological Roots of Medical and Architectural Wonders

Contents

1

Chapter 1: Introduction to the Mythological Connection

Throughout human history, myth and reality have often intertwined, creating a rich tapestry of beliefs, stories, and wonders. This relationship is especially evident in the fields of medicine and architecture, where ancient myths have influenced the design and purpose of structures and the healing practices within them. From the temples of ancient Greece to the towering cathedrals of Europe, these edifices serve as reminders of humanity's quest for understanding and connection with the divine. This book delves into the mythological roots of some of the most fascinating medical and architectural wonders, exploring how these stories have shaped the world we live in today.

The intertwining of myth and reality is not just a relic of the past but continues to shape modern society in profound ways. Myths provide a framework for understanding complex concepts and offer a narrative that connects the physical and metaphysical worlds. In the realm of architecture, mythological themes can be seen in the design of buildings that aim to inspire and evoke a sense of wonder. Similarly, in medicine, the ancient stories of healing gods and miraculous cures continue to influence contemporary practices and symbols.

As we journey through the pages of this book, we will uncover the stories behind some of the most iconic structures and medical traditions. Each chapter will shed light on the mythological roots of these wonders, revealing the intricate connections between belief, design, and healing. By exploring these mythological connections, we gain a deeper appreciation for the ingenuity and creativity of our ancestors, as well as a greater understanding of the cultural and spiritual significance of these wonders.

This exploration is not merely an academic exercise but a celebration of the human spirit and its enduring quest for knowledge, healing, and connection with the divine. Through the lens of mythology, we can see how our ancestors sought to explain the mysteries of the world and how their stories continue to inspire and guide us today. Join us on this journey as we explore the temples of the flesh and the mythological roots of medical and architectural wonders.

2

Chapter 2: The Healing Temples of Asclepius

The ancient Greeks believed in Asclepius, the god of medicine and healing. Temples dedicated to Asclepius, known as Asclepieia, were built throughout Greece and served as centers for medical treatment and spiritual healing. Pilgrims would travel to these temples to seek cures for their ailments, often spending the night in a process called "incubation," where they would sleep in a designated area and hope to receive a healing dream or vision from the god. The priests of Asclepius, known as Asclepiades, would interpret these dreams and prescribe treatments based on their insights, blending religious belief with early medical practices.

The Asclepieia were not merely places of worship but also early hospitals that combined faith with medical knowledge. The temples featured various facilities, including sleeping chambers, baths, and exercise areas, where patients could receive holistic treatments. The healing rituals often involved prayers, sacrifices, and the use of medicinal herbs, showcasing a deep connection between spirituality and medicine. The influence of Asclepius extended beyond Greece, with temples dedicated to him found in other parts of the ancient world, including Rome and Asia Minor.

The mythological stories surrounding Asclepius also played a significant role in shaping the practices within the temples. According to legend,

Asclepius possessed the power to bring the dead back to life, a gift bestowed upon him by the gods. This ability led to his association with healing and the belief that his temples were places where miraculous cures could occur. The presence of sacred snakes within the Asclepieia further symbolized regeneration and healing, as these creatures were believed to carry the god's blessings.

The legacy of the healing temples of Asclepius can still be seen in modern medical practices and symbols. The Rod of Asclepius, a staff with a single serpent coiled around it, remains one of the most enduring emblems of medicine. The holistic approach to healing, combining physical, mental, and spiritual well-being, continues to influence contemporary healthcare, reminding us of the profound impact of ancient myths on the development of medical science.

3

Chapter 3: The Symbolism of the Rod of Asclepius

One of the most enduring symbols of medicine is the Rod of Asclepius, a staff with a single serpent coiled around it. This emblem has its roots in the mythological stories of Asclepius and his association with snakes, which were considered sacred and symbolized healing and regeneration. The serpent's ability to shed its skin and emerge renewed mirrored the process of healing and recovery. Over time, the Rod of Asclepius became a universal symbol for the medical profession, representing the enduring connection between ancient myths and modern medicine.

The origins of the Rod of Asclepius can be traced back to the temples of Asclepius, where snakes were often kept as sacred animals. These serpents were believed to possess healing powers, and their presence within the temples was a testament to the god's influence. Patients seeking treatment at these temples would sometimes encounter the snakes, and their interactions with the creatures were considered part of the healing process. The serpent's symbolism extended beyond physical healing, representing transformation, renewal, and the cycle of life and death.

The Rod of Asclepius has often been confused with the caduceus, a staff with two snakes intertwined around it and topped with wings. The caduceus is traditionally associated with Hermes, the messenger god, and symbolizes commerce and negotiation. Despite this distinction, the caduceus has also become a widely recognized symbol of medicine, particularly in the United States. This conflation of symbols highlights the complex interplay between mythology and symbolism in the medical field.

Today, the Rod of Asclepius serves as a powerful reminder of the ancient roots of medical practice and the enduring influence of mythology on our understanding of healing. The symbol is used by various medical organizations and institutions worldwide, reflecting the universal nature of the quest for health and well-being. The Rod of Asclepius embodies the belief that medicine is not just a science but an art, deeply connected to the spiritual and mythological traditions that have shaped human civilization.

4

Chapter 4: The Egyptian Temples of Imhotep

Imhotep, an architect, physician, and high priest, is one of the most renowned figures in ancient Egyptian history. He is credited with designing the Step Pyramid of Djoser, one of the earliest and most significant architectural achievements in Egypt. Imhotep's contributions to medicine were equally impressive, and he was later deified as a god of healing and wisdom. Temples dedicated to Imhotep were constructed, where people would seek cures for their ailments through prayers, rituals, and consultations with priests. These temples served as centers of learning and healing, blending architectural grandeur with medical advancements.

The Step Pyramid of Djoser, located in the Saqqara necropolis, is a testament to Imhotep's architectural genius. This monumental structure, which served as a burial complex for Pharaoh Djoser, marked a significant departure from the traditional mastaba tombs of the time. The pyramid's innovative design, with its six stacked tiers, symbolized the pharaoh's ascent to the divine realm. Imhotep's ability to harmonize architectural innovation with spiritual symbolism set the stage for future developments in Egyptian architecture.

Imhotep's legacy as a healer is equally remarkable. Ancient texts and inscriptions credit him with a profound knowledge of medicine, including the use of herbs, surgery, and holistic treatments. His deification as a god of healing reflects the high regard in which he was held by the Egyptians. Temples dedicated to Imhotep, known as "House of Life" institutions, became centers of medical learning and practice. Priests and physicians would study medical texts, perform surgeries, and offer treatments, all while invoking Imhotep's divine guidance.

The influence of Imhotep extended beyond Egypt, with his legend and achievements inspiring future generations of architects and healers. The blending of architectural innovation with medical knowledge in the temples of Imhotep demonstrates the interconnectedness of different fields of human endeavor. Imhotep's enduring legacy is a testament to the power of myth and its ability to inspire greatness in both the arts and sciences.

5

Chapter 5: The Altar of Zeus at Pergamon

The Altar of Zeus at Pergamon, an ancient city in modern-day Turkey, is a stunning example of Hellenistic architecture and religious devotion. This monumental structure was dedicated to Zeus, the king of the gods, and adorned with intricate friezes depicting the Gigantomachy, the battle between the gods and giants. The altar's construction and artistic details were deeply influenced by mythological stories, reflecting the city's reverence for the divine and its desire to connect with the gods. The Altar of Zeus stands as a testament to the powerful interplay between myth and architecture in ancient cultures.

The altar, built during the reign of King Eumenes II in the 2nd century BCE, was designed to be a focal point of worship and communal gatherings. Its grand staircase and towering columns created an awe-inspiring presence, drawing worshippers and visitors from far and wide. The friezes that adorned the altar showcased the artistic mastery of the Pergamene sculptors, who brought the mythological battle to life with dynamic and expressive figures. The depiction of the gods' triumph over the giants symbolized the victory of order over chaos, reinforcing the cultural and religious values of the time.

The Altar of Zeus was more than just a religious monument; it was a statement of the city's power and prestige. The construction of such an elaborate structure required significant resources and skilled labor, reflecting the prosperity and ambition of Pergamon. The altar's design incorporated elements of both Greek and Near Eastern architectural traditions, showcasing the cultural exchange and influences that shaped the Hellenistic world. The integration of mythological themes into the architectural design highlighted the deep connection between religion, politics, and art.

The legacy of the Altar of Zeus continues to be felt today, with its influence seen in various architectural and artistic endeavors. The altar's friezes are now housed in the Pergamon Museum in Berlin, where they continue to inspire and captivate visitors. The blending of mythological storytelling with architectural grandeur in the Altar of Zeus serves as a reminder of the enduring power of myths to shape and define human civilization.

6

Chapter 6: The Architecture of Healing in Ancient India

In ancient India, the concept of healing was closely linked to spirituality and architecture. The Vastu Shastra, an ancient Indian science of architecture and design, emphasized the importance of creating harmonious spaces that promoted physical and mental well-being. Temples and healing centers were designed according to these principles, ensuring that the structures aligned with natural elements and cosmic energies. Ayurveda, the traditional system of medicine in India, also drew from mythological stories and incorporated rituals and practices that aimed to restore balance and harmony within the body and mind.

The principles of Vastu Shastra were believed to create environments conducive to healing and spiritual growth. According to this ancient science, the layout and orientation of a building, as well as the materials used in its construction, could influence the health and well-being of its occupants. Temples and healing centers were often built near water bodies or in locations with positive energy, as these factors were thought to enhance the therapeutic effects of the space. The integration of architectural and medicinal knowledge in ancient India highlights the holistic approach to health and well-being that characterized the culture.

Ayurveda, the traditional system of medicine in India, is deeply rooted in the mythological stories of gods and sages. The texts of Ayurveda, such as the Charaka Samhita and Sushruta Samhita, contain references to mythical figures who imparted knowledge of healing practices to humanity. These stories emphasized the divine origin of medical knowledge and reinforced the belief that health and spirituality were interconnected. Ayurvedic treatments often involved rituals, prayers, and the use of medicinal herbs, reflecting the influence of mythology on the practice of medicine.

The legacy of ancient Indian architecture and healing practices continues to influence modern approaches to health and wellness. The principles of Vastu Shastra have found renewed interest in contemporary architecture and interior design, with an emphasis on creating spaces that promote well-being. Ayurveda has gained global recognition for its holistic approach to health, with many people seeking its treatments for various ailments. The enduring connection between mythology, architecture, and medicine in ancient India serves as a testament to the profound wisdom and ingenuity of the culture.

Chapter 7: The Influence of Roman Mythology on Architecture

The ancient Romans were heavily influenced by Greek mythology, and this is evident in their architectural achievements. Temples, amphitheaters, and public buildings were often dedicated to Roman gods and goddesses, reflecting the empire's reverence for the divine. The Pantheon, a magnificent temple in Rome, was dedicated to all the gods and featured a grand dome that symbolized the heavens. The myths of gods like Jupiter, Mars, and Venus were intricately woven into the design and purpose of these structures, creating a lasting legacy of mythological influence on Roman architecture.

Roman temples were designed to be grand and awe-inspiring, with towering columns, intricate carvings, and expansive courtyards. The architectural style often incorporated elements from Greek and Etruscan traditions, creating a unique blend that defined Roman architecture. The use of mythological themes in the design of temples served to reinforce the cultural and religious beliefs of the time. For example, the Temple of Jupiter Optimus Maximus on Capitoline Hill was a symbol of Rome's power and divine favor, reflecting the importance of the chief god in Roman religion.

Amphitheaters, such as the Colosseum, were another significant aspect of Roman architecture influenced by mythology. These grand structures were designed for public spectacles, including gladiatorial contests, theatrical performances, and mythological reenactments. The Colosseum, in particular, showcased the Romans' engineering prowess and their fascination with mythological stories. The elaborate performances often depicted scenes from mythology, bringing the stories of gods and heroes to life for the audience.

The influence of Roman mythology on architecture extended beyond religious and entertainment structures. Public buildings, such as basilicas and baths, often featured mythological motifs and statues. The Baths of Caracalla, for example, were adorned with mosaics and sculptures depicting gods and mythological scenes, reflecting the integration of art, architecture, and mythology in Roman society. The enduring legacy of Roman architecture serves as a testament to the cultural and religious significance of mythology in shaping the built environment.

8

Chapter 8: The Medical Miracles of the Medieval Cathedrals

During the medieval period, cathedrals served not only as places of worship but also as centers of healing and medical knowledge. The construction of these grand structures was often inspired by religious devotion and mythological stories. Pilgrims would visit cathedrals to seek miraculous cures from relics of saints and martyrs, believing in the divine power of these holy objects. The architecture of the cathedrals, with their soaring spires and intricate stained glass windows, was designed to inspire awe and create a sense of connection with the divine. This blend of faith, myth, and architectural wonder contributed to the cathedrals' reputation as places of healing and hope.

The belief in the healing power of relics was a central aspect of medieval Christianity. Saints and martyrs were venerated for their miraculous deeds, and their remains or personal items were believed to possess divine healing properties. Cathedrals often housed these relics in elaborate reliquaries, attracting pilgrims from far and wide. The faithful would pray before the relics, seeking cures for their ailments and attributing their recovery to the intercession of the saints. The connection between relics and healing reinforced the belief in the divine presence within the cathedrals.

The architectural design of medieval cathedrals played a significant role in creating an atmosphere of reverence and spiritual connection. The use of pointed arches, ribbed vaults, and flying buttresses allowed for the construction of towering structures with expansive interiors. The stained glass windows, depicting scenes from the Bible and the lives of saints, filled the cathedrals with colorful light, creating a sense of divine illumination. The architectural grandeur and artistic details of the cathedrals reflected the devotion and craftsmanship of the medieval builders, who sought to create spaces that inspired faith and awe.

The legacy of medieval cathedrals as centers of healing continues to be felt today. Many cathedrals still attract pilgrims seeking spiritual solace and physical healing. The architectural and artistic achievements of these structures continue to inspire awe and admiration, serving as reminders of the profound connection between faith, architecture, and healing. The medieval cathedrals stand as enduring symbols of the human quest for divine intervention and the belief in the miraculous.

9

Chapter 9: The Healing Powers of the Native American Medicine Wheel

The Native American Medicine Wheel is a powerful symbol of healing and spirituality, deeply rooted in the mythology and traditions of various indigenous tribes. The circular design of the Medicine Wheel represents the interconnectedness of all life and the cyclical nature of existence. Each direction on the wheel is associated with specific elements, animals, and spiritual meanings, guiding individuals on their journey to physical, emotional, and spiritual well-being. The construction of Medicine Wheel structures, often made of stones arranged in a circular pattern, served as sacred spaces for healing ceremonies and rituals, reflecting the profound connection between mythology and architecture in Native American cultures.

The concept of the Medicine Wheel varies among different Native American tribes, but its core symbolism remains consistent. The four cardinal directions—north, south, east, and west—are each associated with specific elements, such as air, water, fire, and earth, as well as animals and colors. These associations provide a framework for understanding the natural world and the individual's place within it. The Medicine Wheel serves as a guide for personal growth and healing, helping individuals to balance their physical, emotional, and spiritual aspects.

Healing ceremonies and rituals conducted within the Medicine Wheel often involve prayers, chants, and the use of sacred herbs and objects. These practices are believed to harness the spiritual energies of the wheel, promoting healing and harmony. The circular design of the Medicine Wheel symbolizes the cyclical nature of life, death, and rebirth, reflecting the belief in the interconnectedness of all living beings. The construction of Medicine Wheel structures in sacred spaces further enhances their spiritual significance, providing a tangible representation of the tribe's mythology and beliefs.

The influence of the Native American Medicine Wheel continues to be felt in contemporary healing practices and spiritual teachings. The principles of balance, harmony, and interconnectedness resonate with many people seeking holistic approaches to health and well-being. The Medicine Wheel serves as a reminder of the rich cultural heritage and spiritual wisdom of Native American tribes, offering valuable insights into the healing power of mythology and sacred architecture.

10

Chapter 10: The Sacred Architecture of the Mayan Civilization

The Mayan civilization, known for its advanced knowledge of astronomy, mathematics, and architecture, created some of the most remarkable structures in the ancient world. Temples, pyramids, and observatories were built to honor their gods and track celestial events. Mythology played a central role in the design and purpose of these structures, with the stories of gods like Kukulkan and Itzamna intricately woven into their creation. The Mayan temples were not only places of worship but also centers of healing, where rituals and ceremonies were performed to seek divine intervention and restore balance to the community.

The Temple of Kukulkan, also known as El Castillo, at the archaeological site of Chichen Itza, is one of the most iconic examples of Mayan architecture. This pyramid-shaped temple is aligned with the movements of the sun and features a remarkable phenomenon during the equinoxes, when the setting sun creates the illusion of a serpent descending the pyramid's steps. This alignment with celestial events reflects the Mayans' deep understanding of astronomy and their belief in the connection between the heavens and the earth. The Temple of Kukulkan served as a focal point for religious ceremonies and rituals, where offerings were made to the god Kukulkan, the feathered serpent.

The Mayan civilization also built observatories, such as the Caracol at Chichen Itza, to study the stars and track celestial events. These structures were designed with precise alignments to observe astronomical phenomena, aiding the Mayans in creating accurate calendars and predicting celestial events. The integration of mythological stories with astronomical observations highlights the holistic approach of the Mayans, where science and spirituality were intertwined. The knowledge gained from these observations was used to plan agricultural activities, religious ceremonies, and community events, reflecting the practical applications of their mythological beliefs.

Healing rituals and ceremonies were an essential aspect of Mayan culture, often conducted in temples and sacred spaces. The Mayans believed that illness and disease were caused by imbalances in the natural and spiritual world, and sought to restore harmony through rituals, prayers, and offerings. Temples dedicated to gods of healing, such as Ix Chel, the goddess of medicine and childbirth, provided a space for these practices. The blending of architecture, mythology, and healing in Mayan civilization demonstrates the interconnectedness of their cultural and spiritual beliefs.

The legacy of Mayan architecture and mythology continues to inspire and captivate people around the world. The precision and grandeur of their structures, combined with their rich mythological traditions, offer valuable insights into the ingenuity and spirituality of the Mayan civilization. The sacred architecture of the Mayans serves as a testament to their advanced knowledge and deep connection with the natural and celestial worlds.

11

Chapter 11: The Renaissance Revival of Mythological Themes

The Renaissance period saw a revival of interest in classical mythology and its influence on art, architecture, and medicine. Architects and artists drew inspiration from ancient Greek and Roman myths, incorporating these themes into their work. The construction of grand structures like St. Peter's Basilica in Vatican City and the Villa Rotonda in Italy reflected a renewed appreciation for mythological stories and their symbolic meanings. Medical practitioners of the Renaissance also looked to the past, studying ancient texts and incorporating mythological references into their treatments and cures, blending science and mythology in their pursuit of knowledge.

St. Peter's Basilica, one of the most famous architectural achievements of the Renaissance, exemplifies the revival of classical themes. Designed by renowned architects such as Bramante, Michelangelo, and Bernini, the basilica features a grand dome and intricate details that draw from classical Roman and Greek architecture. The use of mythological motifs in the design, such as the depiction of angels and saints, reflects the Renaissance fascination with the divine and the symbolic meanings of these stories. The basilica's construction was a monumental undertaking that showcased the artistic and architectural advancements of the period.

The Villa Rotonda, designed by the architect Andrea Palladio, is another iconic example of Renaissance architecture inspired by classical mythology. The villa's symmetrical design, with its central dome and porticos, draws from the principles of classical Greek and Roman architecture. Palladio's work was deeply influenced by the writings of the Roman architect Vitruvius, whose texts on architecture emphasized the importance of harmony, proportion, and beauty. The integration of mythological themes into the design of the Villa Rotonda reflects the Renaissance ideal of combining artistic beauty with intellectual and spiritual significance.

In the field of medicine, Renaissance practitioners looked to ancient texts and mythological stories for guidance and inspiration. The study of ancient Greek and Roman medical works, such as those by Hippocrates and Galen, informed the practices of Renaissance physicians. Mythological references were often used to explain medical concepts and treatments, blending scientific knowledge with symbolic meanings. The use of mythological imagery in medical texts and illustrations helped to convey complex ideas and reinforce the connection between health, spirituality, and the natural world.

The Renaissance revival of mythological themes in art, architecture, and medicine highlights the enduring influence of ancient myths on human creativity and knowledge. The integration of classical stories into the cultural and intellectual achievements of the period reflects the Renaissance belief in the timeless value of these myths. The legacy of this revival continues to inspire contemporary artists, architects, and medical practitioners, reminding us of the profound impact of mythology on the development of human civilization.

12

Chapter 12: The Modern Legacy of Mythological Influences

In the modern era, the influence of mythology on medicine and architecture continues to be felt. Contemporary architects and designers often draw inspiration from ancient myths and symbols, creating structures that blend aesthetic beauty with deeper meanings. The use of mythological motifs in medical institutions, such as the incorporation of the Rod of Asclepius in hospital logos, serves as a reminder of the enduring connection between myth and healing. As we move forward, the lessons of the past and the stories of our ancestors will continue to shape our understanding of the world and inspire us to create wonders of our own.

Modern architecture often incorporates mythological themes to create a sense of connection with history and culture. Buildings such as museums, theaters, and public spaces are designed with symbolic elements that evoke ancient myths and stories. The use of classical motifs, such as columns and friezes, reflects a reverence for the architectural achievements of the past. These designs not only enhance the aesthetic appeal of the structures but also create a sense of continuity with the cultural heritage of humanity. The blending of modern technology with mythological symbolism highlights the enduring relevance of these stories in contemporary society.

In the field of medicine, the influence of mythology is evident in various symbols and practices. The Rod of Asclepius, for example, remains a prominent emblem of the medical profession, symbolizing the healing power and wisdom associated with the ancient god. Medical institutions often incorporate mythological imagery in their logos, designs, and educational materials, reinforcing the connection between ancient beliefs and modern practices. The holistic approach to healing, which emphasizes the integration of physical, mental, and spiritual well-being, reflects the enduring influence of mythological concepts on contemporary healthcare.

The lessons of ancient myths continue to inspire and guide modern innovations in medicine and architecture. The emphasis on harmony, balance, and beauty in architectural design reflects the principles of classical mythology, while the incorporation of holistic healing practices in medicine draws from ancient beliefs in the interconnectedness of all aspects of life. By studying and appreciating the mythological roots of these fields, we gain valuable insights into the timeless wisdom of our ancestors and the enduring relevance of their stories.

The modern legacy of mythological influences serves as a reminder of the profound impact of these stories on human civilization. As we continue to explore and innovate, the myths of the past will remain a source of inspiration, guiding us in our quest for knowledge, healing, and beauty. The temples of the flesh, both ancient and modern, stand as enduring symbols of the human spirit and its unending pursuit of connection with the divine.

13

Chapter 13: The Mythological Influence on Chinese Medicine and Architecture

Chinese mythology and traditional medicine are deeply intertwined, with ancient stories influencing healing practices and architectural designs. The Taoist philosophy, which emphasizes harmony between humans and nature, has played a significant role in shaping Chinese medicine and architecture. Temples and healing centers were often constructed according to Feng Shui principles, which aimed to create balanced and harmonious environments. The mythological tales of deities like Shen Nong, the divine farmer and healer, and the Eight Immortals have left an indelible mark on Chinese culture, inspiring both medical practices and architectural wonders.

Shen Nong, a legendary figure in Chinese mythology, is credited with discovering medicinal herbs and teaching agriculture. Often depicted as a deity with a transparent stomach, Shen Nong could see the effects of herbs on his body, allowing him to identify their medicinal properties. His contributions to Chinese medicine are chronicled in the ancient text "Shen Nong Ben Cao Jing," which remains a cornerstone of traditional Chinese medicine. Temples dedicated to Shen Nong and other healing deities became centers of medical knowledge and spiritual healing, where practitioners would study and perform treatments based on ancient wisdom.

Feng Shui, the ancient Chinese art of geomancy, emphasizes the importance of creating harmonious living spaces that promote health and well-being. The principles of Feng Shui are rooted in Taoist philosophy and Chinese mythology, which teach that the arrangement of physical spaces can influence the flow of energy, or "Qi." Temples, homes, and healing centers were designed with careful consideration of Feng Shui principles, ensuring that they aligned with natural elements and cosmic forces. The integration of mythological concepts into architectural design reflects the holistic approach of Chinese culture, where health and spirituality are deeply connected.

The influence of Chinese mythology and traditional medicine continues to be felt in modern practices. The use of acupuncture, herbal medicine, and Qi Gong exercises, all rooted in ancient beliefs, remains a vital part of Chinese healthcare. The architectural principles of Feng Shui are also widely adopted in contemporary design, promoting harmony and balance in living spaces. The enduring legacy of Chinese mythology and traditional medicine serves as a testament to the cultural richness and spiritual wisdom of the civilization.

14

Chapter 14: The Mythological Foundations of Islamic Medicine and Architecture

Islamic culture and mythology have had a profound impact on the development of medicine and architecture in the Islamic world. The rich tapestry of Islamic myths, stories, and beliefs has inspired both the healing practices and the architectural marvels that define the region. The Islamic Golden Age, a period of flourishing scientific, cultural, and intellectual achievements, saw significant advancements in medicine, many of which were influenced by mythological and religious concepts. Similarly, the architectural splendor of Islamic mosques, palaces, and gardens reflects the deep connection between faith, mythology, and design.

The Islamic Golden Age, which spanned from the 8th to the 14th centuries, was a time of remarkable progress in various fields, including medicine. Scholars and physicians, such as Avicenna (Ibn Sina) and Al-Razi, made significant contributions to medical knowledge, drawing inspiration from both Greek and Islamic traditions. Avicenna's "Canon of Medicine," a comprehensive medical encyclopedia, became a foundational text in both the Islamic world and Europe. The integration of mythological and religious concepts into medical practices, such as the belief in divine healing and the use of prophetic medicine, highlights the enduring influence of Islamic mythology on the field.

Islamic architecture, renowned for its beauty and intricacy, is deeply rooted in the cultural and mythological traditions of the region. The design of mosques, palaces, and gardens often incorporates symbolic elements that reflect Islamic beliefs and stories. The use of geometric patterns, arabesques, and calligraphy in architectural decoration serves as a visual representation of the divine order and the unity of creation. The Alhambra in Spain and the Great Mosque of Cordoba are prime examples of the architectural splendor inspired by Islamic mythology and faith.

Gardens, known as "paradise gardens," are another significant aspect of Islamic architecture influenced by mythology. These gardens were designed to evoke the Quranic description of paradise, with lush vegetation, flowing water, and shaded walkways creating a serene and harmonious environment. The concept of the garden as a reflection of divine paradise underscores the connection between Islamic mythology and architectural design, emphasizing the importance of harmony, beauty, and spirituality.

The legacy of Islamic mythology and its influence on medicine and architecture continues to inspire contemporary practices. The holistic approach to health, which considers the physical, mental, and spiritual well-being of individuals, remains an essential aspect of Islamic medicine. The architectural principles and symbolic elements rooted in Islamic mythology continue to shape the design of modern buildings, reflecting the enduring cultural and spiritual heritage of the Islamic world.

15

Chapter 15: The Mythological Inspirations of Japanese Medicine and Architecture

Japanese mythology, with its rich pantheon of gods, spirits, and folklore, has profoundly influenced the country's approach to medicine and architecture. The Shinto religion, which emphasizes the worship of kami (spiritual beings) and the harmony between humans and nature, plays a central role in shaping Japanese cultural practices. Temples, shrines, and healing centers are often designed according to Shinto principles, reflecting the deep connection between spirituality, health, and the natural world. The stories of deities like Susanoo, the storm god, and Omoikane, the deity of wisdom, have left an indelible mark on Japanese traditions, inspiring both medical practices and architectural marvels.

Shinto shrines, which serve as places of worship and community gatherings, are designed to harmonize with their natural surroundings. The architectural style of these shrines often features wooden structures, thatched roofs, and natural materials, reflecting the belief in the sacredness of nature. The layout of the shrines, with their torii gates, purification rituals, and sacred spaces, emphasizes the importance of spiritual cleansing and connection with the kami. Healing rituals conducted at these shrines often involve prayers, offerings, and the use of sacred water and herbs, reflecting the influence of mythology on medical practices.

Traditional Japanese medicine, known as Kampo, has its roots in both Chinese medicine and indigenous Japanese beliefs. Kampo incorporates the use of herbal remedies, acupuncture, and moxibustion, drawing from ancient texts and mythological stories. The belief in the interconnectedness of the body, mind, and spirit is a central aspect of Kampo, reflecting the holistic approach to health and healing in Japanese culture. The influence of Shinto and Buddhist mythology on Kampo practices highlights the integration of spiritual and medicinal knowledge.

Japanese architecture, particularly the design of temples and gardens, is deeply influenced by mythological and religious concepts. Zen gardens, with their carefully arranged rocks, sand, and vegetation, are designed to evoke a sense of tranquility and reflection, mirroring the principles of Zen Buddhism. The use of natural elements and minimalist design in Japanese architecture reflects the cultural emphasis on harmony, simplicity, and connection with nature. The stories and symbolism embedded in these structures serve as a testament to the profound impact of mythology on Japanese art and design.

The legacy of Japanese mythology and its influence on medicine and architecture continues to inspire contemporary practices. The holistic approach to health and the emphasis on harmony with nature remain integral aspects of Japanese culture. The architectural principles and symbolic elements rooted in Shinto and Buddhist traditions continue to shape the design of modern buildings and spaces. The enduring connection between mythology, medicine, and architecture in Japan serves as a testament to the cultural richness and spiritual wisdom of the civilization.

Book Description

"Temples of the Flesh: Exploring the Mythological Roots of Medical and Architectural Wonders" is a captivating journey through the rich tapestry of myths, stories, and beliefs that have shaped the fields of medicine and architecture throughout human history. From the healing temples of ancient Greece to the grand cathedrals of medieval Europe, this book delves into the profound connection between myth and reality, revealing how ancient stories have influenced the design and purpose of some of the most iconic structures and medical practices.

Through twelve insightful chapters, readers will explore the mythological roots of medical and architectural wonders from various cultures, including the Asclepieia of ancient Greece, the Egyptian temples of Imhotep, the Mayan pyramids, and the Islamic gardens of paradise. Each chapter sheds light on the intricate connections between belief, design, and healing, offering a deeper appreciation for the ingenuity and creativity of our ancestors.

With additional chapters on Chinese, Islamic, and Japanese traditions, "Temples of the Flesh" provides a comprehensive exploration of the enduring influence of mythology on contemporary practices. The book celebrates the timeless wisdom of our ancestors and their unending quest for knowledge, healing, and connection with the divine. Whether you are a history enthusiast, a lover of mythology, or simply curious about the world, this book offers a fascinating and enlightening journey through the mythological roots of our most cherished medical and architectural achievements.